Hands-On Math and Literature with MathStart®

Level 2

by Don Balka
and Richard Callan

Printed in the United States of America.

This book is printed on recycled paper.

Order Number 2-196
ISBN 978-1-58324-238-4

H I J K L 18 17 16 15 14

395 Main Street
Rowley, MA 01969
www.didax.com

CONTENTS

ABOUT THE AUTHORS

Don S. Balka, Ph.D., is a noted mathematics educator who has presented numerous workshops on the use of mathematics manipulatives with elementary, middle and high school students at national and regional conferences of the National Council of Teachers of Mathematics, at state conferences and at in-service training for school districts throughout the world. He is a former junior high and high school mathematics teacher, and is currently a Professor in the Mathematics Department at Saint Mary's College, Notre Dame, Indiana.

Richard J. Callan has been a public school teacher in Indiana for 25 years and holds his BS and MS from Indiana University. He conducts staff development workshops and makes presentations on children's literature, assessment and manipulatives. In 1995, he received the Presidential Award for Elementary School Mathematics and appears in *Who's Who in America* and *Who's Who Among America's Teachers*. He was a contributing author to the *Indiana Mathematics Proficiency Guides* in 1991 and 1997. He also was Program Chair for the NCTM Central Regional Conference, January of 2003. Rick also was the mathematics representative for the SEPA (Society of Elementary Presidential Awardees) and co-authored the books *Math, Literature and Unifix* and *Math, Literature and Manipulatives* with Dr. Don Balka.

INTRODUCTION

Integrating children's literature into your mathematics program can be a fresh, enriching experience for you as a teacher and for your children in learning mathematics, for appreciating mathematics in varied settings, and for understanding mathematics in a non-threatening, inviting environment. In *Hands-On Math and Literature with MathStart,* more fun and often challenging activities are provided to supplement those in books.

With his innovative MathStart series that includes books in three different levels, well-known children's author Stuart Murphy gives children a unique way to understand and develop the mathematics content. Each core topic selected by Murphy for his books correlates with the National Council of Teachers of Mathematics Principles and Standards for School Mathematics (2000). In many cases, the books are appropriate matches for local and state standards. The levels are by ages:

- Level 1: Ages 3 and up
- Level 2: Ages 6 and up
- Level 3: Ages 7 and up

Each level of the MathStart series examines various core topics in 21 different books. The readability of each book differs and should not be the determining factor in using the book for a specific grade level. Besides regular classroom students, special needs and ESL students will benefit from their teachers using books in the series as instructional tools or as reinforcements for concepts taught in class. Extensions of the mathematics presented in several of the books are appropriate for middle and high school students.

For each book in the MathStart series, *Hands-On Math and Literature with MathStart* presents the following pertinent information for teachers:

- Title
- Story Summary
- Grade Level
- Concepts or Skills
- Objectives
- Materials Needed

- Activities
- Writing Activities
- Internet Sites
- Assessment Ideas

Books in MathStart, Level 2, cover the following topics: Adding, comparing sizes, directions, recognizing shapes, doubling numbers, counting, opposites, ordinals, understanding capacity, hours, counting on, comparing amounts, counting by 5s and 10s, comparing weights, odd and even numbers, subtracting one, ordering numbers, matching, sequencing, matching sets and sorting.

Besides the activities Murphy suggests at the end of each book, additional activities for other mathematics concepts are provided for teachers to use to expand or extend their students' mathematical learning and understanding. Teachers will be able to use these activities to develop their own lessons or thematic units of mathematics study.

Internet sites have been listed with some book entries for teacher's perusal. Some sites are inclusive with other core topics, while other sites are specific for one topic or book.

Writing or communication activities have been presented for students to think, talk, or draw about in a class or small group situation. Some of the writing prompts will provide teachers with feedback as to whether students have understood the mathematics presented. Other writing prompts provide students with opportunities to expand their thoughts and understanding of the mathematics presented in the stories.

The assessment component will let teachers measure the understanding of the mathematics using a pencil and paper task, a performance task with manipulatives, or a writing assignment. Some Internet sites will allow teachers to assess students' understanding also.

Children's literature and appropriate activities with manipulatives can be an inviting experience for children to learn and understand mathematics. By using manipulatives in the classroom, children will be able to understand mathematical information, develop mathematical concepts beyond conventional classroom settings, independently learn and understand mathematical concepts, rejuvenate creative thinking, have an appreciation for reading, and have a focal point on problem solving strategies and using connections to everyday living.

NCTM CORRELATION

	Number & Operations	Algebra	Geometry	Measurement	Data Analysis & Probability
100 Days of Cool	√			√	
Animals on Board	√				
The Best Vacation Ever					√
Bigger, Better, Best!				√	
Captain Invincible and the Space Shapes			√		
Coyotes All Around	√				
Elevator Magic	√				
A Fair Bear Share	√				
Get Up and Go!	√			√	
Give Me Half!	√				
Let's Fly a Kite			√		
Mall Mania	√				√
More or Less	√	√			
Pepper's Journal				√	√
Probably Pistachio					√
Racing Around				√	
Same Old Horse					√
Spunky Monkeys on Parade	√				
The Sundae Scoop					√
Super Sand Castle Saturday				√	
Tally O'Malley					√

100 DAYS OF COOL

Story Summary

Maggie incorrectly hears her new teacher say that the class was going to celebrate 100 days of "cool" rather than 100 days of "school." On the first day of school, Maggie, Nathan, Yoshi and Scott dressed wildly, only to find the mistake Maggie had made. Although surprised, Mrs. Lopez likes the idea and asks the children to keep it going for 99 more days. Silly things occur over the 100 days, until finally the class is able to celebrate 100 days of cool with a special party.

New York: Harper Collins Publishers, 2004   ISBN: 0-06-000121-6

Concepts or Skills

- Number sense
- Number line
- Time
- Fractions
- Decimals

Objectives

- Construct a number line
- Determine fractional or decimal parts of 100
- Determine 100 units of time (seconds, minutes, hours, days)

Materials Needed

- Unifix Cubes
- 100 Square Grid, page 52
- Monthly calendars
- 100 pennies, nickels, dimes and quarters
- Small cereal or candies

Activity 1

Copy and distribute calendars that will incorporate the first 100 days of school. Depending on the starting dates for school, the 100th day of school should occur in January or February in most places. To begin, have students cross out days when there will be no school or cross these out before distributing. On each day of class, have students write the day of class (1 - one, 2 - two, 3 - three, 4 - four, …, 100 - one hundred).

Activity 2

Divide the class into small groups and distribute 100 Unifix Cubes to each group. Have students connect the cubes into a long bar on the floor. Using the cubes as a measuring bar, have students find each other's height. Record the results and discuss class findings.

Have students find objects in the class that are approximately 100 cubes in length.

Activity 3

Copy and distribute the 100 Square Grid. For each day of school, have students color one square on the grid. The same color should be used for each square in group of ten days. Have students change colors for each new group of ten days.

Discuss the decimal fractions 1/10, 2/10, …, 10/10; 1/100, 2/100, …, 100/100 on the appropriate days. Discuss writing the decimal fractions as decimals .1, .2, .3, …, 1; .01, .02, .03, …, 1

Activity 4

With 100 as the "target" number, talk about time units in terms of 100.

- 100 seconds
- 100 minutes. If you started counting minutes at 8 a.m. on a particular day, then the 100th minute would occur at 9:40 a.m.
- 100 hours is slightly more than 4 days. If you started counting hours at 8 a.m. on a particular day, then the 100th hour would be at noon on the fifth day.

Activity 5

Distribute bags of coins to groups. Create bags so that there are ten coins of each denomination (penny, nickel, dime, quarter). Have students carefully stack each denomination of coin and then measure the height of the stack of ten. Multiplying by 10 would then give the height for a stack of 100.

Have students make a bar graph showing the results of their measurements. Here are approximate heights and lengths for each denomination.

- 100 Pennies: 15 cm   190 cm
- 100 Nickels: 20 cm   212 cm
- 100 Dimes: 12 cm   180 cm
- 100 Quarters: 16 cm   241 cm

Have students determine the amount of money in dollars for each denomination.

Using the information from above, have students determine the height and length for $100 for each denomination. Using the Internet, have students go to www.usmint.gov to find the exact height of each coin. Have students place the coins side-by-side and find the total length for 100 coins.

Activity 6

Using cereal or candy, such as Cheerios, have students count out 100 and place them in a bowl. Have students estimate how many Cheerios are in a box, and then count them. For upper grades, use different size boxes to determine the best buy.

Activity 7

Using any operation, have students create different number sentences that result in an answer of 100. List their sentences on chart paper and discuss the different types of sentences they have created.

Activity 8

Have a clear plastic jar displayed in the classroom. On the first day of school, place one penny in the jar. As each day goes by, add one penny. Depending on grade level, have students exchange the pennies for other coins when possible. Continue adding pennies until the 100th day is reached.

For upper grades, place one penny (or plastic penny) in the jar on the first day, 2 pennies on the second day, and so on until the 100th day. Have students estimate how much money will be in the jar after 100 days. As a problem of the week, have students determine how many pennies or how much money is in the jar each week.

For 100 days, there will be 5050 pennies or $50.50. A big jar is needed for this project. To determine the number of pennies on any particular day, use the following formula:

$$\text{amount of money} = \frac{N(N + 1)}{2}$$

where N = number of days. For example, on the 3rd day, there will be 6¢ in the jar.
Since N = 3, then the amount of money is 3(4)/2 = 6¢.

Writing and Communicating

Have students create their own stories about 100.

Have students write an answer to "How can numbers be cool?"

Have students categorize different ways to write 100.

Assessment

Shade small 100 Square Grids and have students determine the fractional part or decimal part shaded.

Have students shade a specified fractional part or decimal part of a 100 Square Grid.

Give students 10 objects that can be stacked or placed in a row. Have them determine the approximate length of 100 objects.

Internet Links

www.usmint.gov

www.coolmath.com

ANIMALS ON BOARD

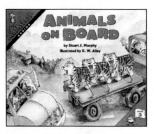

Story Summary

Jill, the truck driver, has a special load. Driving along, she encounters several trucks that pass her with different loads of animal models. There are two sets of animals that come by on two different trucks. In the story, the sum of the numbers of each animal is displayed. Finally, Jill reaches her destination. She has been hauling the top for an animal merry-go-round.

New York: Harper Collins Publishers, 1998   ISBN: 0-06027442-5

Concepts or Skills

- Basic addition facts

Objectives

- Add two single digit numbers
- Model a basic addition fact

Materials Needed

- Animal crackers, Teddy Grahams, Goldfish Colors Crackers, or Scooby-Doo Crackers
- Library books about animals
- Unifix Cubes
- Totaling 10 Cards, page 53

Activity 1

Give each student some animal crackers that are named in the materials list. Have students model addition sentences using the crackers. As the teacher reads the book aloud, have children model the equations that are presented in the book by showing each addend and combining to find the sum.

Have children model their own number sentences and find the sums.

Activity 2

The story involves tigers, swans, frogs, horses and pandas. Have children list one mathematical fact about each animal, for example, number of legs, number of eyes. Share the facts with the entire class.

Activity 3

Play the solitaire game, Totaling 10. Make copies of the cards on cardstock and cut them apart. Cut off the upper left hand corner on each card. Give each child a deck of cards and have them shuffle the cards.

Lay the cards out in two rows of three, with the numbers showing. Keep placing cards on top of the piles in each row. The cards are dealt left to right, starting with the top row. This row will eventually have 4 cards per pile, while the bottom row will have 3 cards per pile when all the cards are dealt.

5 2	3 2	0 0

6 2	3 0	1 1

The student picks up two cards so that the sum of the numbers is 10. In the above array, for example, a student could pick up the $\begin{smallmatrix}5\\2\end{smallmatrix}$ and $\begin{smallmatrix}3\\0\end{smallmatrix}$ cards, then the $\begin{smallmatrix}6\\2\end{smallmatrix}$ and $\begin{smallmatrix}1\\1\end{smallmatrix}$ cards. A student must pick up two cards at all times until there is only one card remaining.

When all of the cards in a pile have been used up, the student moves a card from another pile to fill the space. Keep two rows of 3 going until there are only a few cards left.

If a student can't find two cards to equal the sum of 10, he may "cheat." This means that a student will need to arrange the cards in the piles until play can resume.

The 21st card is the check card. This last card, indicating a player has won, will always sum to 6. If the 21st card doesn't sum to 6, then the player has added incorrectly somewhere in the game. Students should crisscross cards as they make 10 so that it is easier to see what has been added incorrectly.

Activity 4

Write an addition sentence such as 3 + 2 = on the board. Have children use Unifix Cubes to model the sentence and find the sum.

Writing and Communicating

Give children the prompt "If I know how to add, then I can …" Assist them in responding.

Give children the prompt "I want to add but my calculator exploded. What can I do?" Assist them in responding.

Assessment

Give children a selected set of animal crackers and have them model various number sentences.

Present children with an abbreviated array of cards from Totaling 10 and have them find two cards that produce a sum of 10.

Give children a selected set of Unifix Cubes and have them model various number sentences.

Internet Link

www.edu4kids.com

Notes:

THE BEST VACATION EVER

Story Summary

A little girl and her family, mom, dad, brother Charlie, grandma, and her cat, Fluffer, are very busy. They need a vacation, but do not know where to go. The girl uses data collection and problem solving to determine where her family should go on vacation...the back yard.

New York: Harper Collins Publishers, 1997   ISBN: 0-06-026766-6

Concepts or Skills

- Data collection
- Graphing

Objectives

- Count to a specified number
- Order a set of objects from shortest to longest
- Order a set of objects from smallest to largest

Materials Needed

- Unifix Cubes
- Grid paper
- Manila paper
- Pizza Toppings Circle, page 54

Activity 1

As a homework assignment, have each student make a list of questions that they might ask their family members about where to take a vacation. Here are some examples:

- Would you like to spend your vacation in a warm or cool place?
- Would you like to spend your vacation near water?
- Would you like to spend your vacation out of state?
- Would you like to spend your vacation visiting relatives?
- Would you like to spend your vacation at home doing things that are close to home?

Students should be encouraged to develop their own questions.

When the data has been collected, students should create a table to display their findings for the whole class. Provide manila paper for students to construct the table to display on bulletin boards.

Activity 2

Have students work in cooperative groups. Let them collect data on the following topics, asking members of their class one of the following questions.

- What is your favorite candy bar?
- What is your favorite television show?
- What is your favorite school subject?
- What is your favorite fast food restaurant?
- What is your favorite cola?
- What is your favorite school lunch?
- In what month were you born? (In what season were you born?)

After completing the data collection, have students decide how to display the data. Depending on grade level, students can construct a variety of graphs on graph paper or using the computer.

Activity 3

Make a copy of the Pizza Toppings Circle. Have children take small sticky notes with their names on them and place them on the area of the circle representing their favorite topping. Discuss the findings. Have students create a different type of graph to illustrate the findings.

Repeat the activity with any of the following: ice cream, ice cream toppings, favorite fruit, favorite cereal.

Activity 4

Write headings for a bar graph on bulletin board paper or on the chalkboard:

Which is the best place to take a vacation?

Disney World	Camping	Beach	Stay at Home	Other

Have students take a small sticky note with their name on it and put it on the column for their favorite vacation spot. Discuss the graph. Have students write about the findings.

Activity 5

To help students understand tally marks, give each student a popsicle stick. On bulletin board paper, write any of the following titles with the corresponding choices.

Have students glue their stick under any of the entrees that are listed on the paper. Note to students that the fifth stick crosses diagonally over four vertical sticks. Discuss why this procedure helps in reading the information shown.

- My Favorite Cake: chocolate, lemon, white, cherry, other
- My Favorite Homemade Cookie: chocolate chip, oatmeal, sugar, m&m's cookies, other

- My Favorite Pie: cherry, apple, strawberry, chocolate, other
- My Favorite Hamburger Topping: ketchup, mustard, onions, cheese, other
- My Favorite Supermarket Cookie: Oreo, Chips Ahoy, other
- My Favorite Subject: reading, spelling, mathematics, social studies, science, health, music, physical education, art
- My Favorite Sport: football, basketball, baseball, soccer, golf, volleyball, other
- My Favorite Drink: water, cola, tea, juice, other
- Do you have a library card? Yes or No
- Do you write with your left hand or right hand? Left or Right
- When I grow up I want to be a: doctor, teacher, lawyer, nurse, professional athlete, artist, movie star, rock star, fireperson, policeperson, business owner, other
- My Hair Color: brown, black, blonde, red, other
- My Favorite Season: fall, winter, spring, summer

Activity 6

Have students keep a log of all the books they read in one month. At the end of a month, tally the number of books read by each student. Make a line graph, where the horizontal axis is labeled with students' names and the vertical axis is labeled with the number of books read. Discuss the line graph.

Writing and Communicating

Have students write about or discuss how collecting data and graphing data help us understand answers to specific questions.

Assessment

Have students construct a specified graph from a given table of data.

Have students interpret a given graph.

Notes:

BIGGER, BETTER, BEST!

Story Summary

Jenny and her brother Jeff were always arguing about who had the bigger or better items. Jill, the younger sister, would cover her head and put her fingers in her ears when they would start the arguments. Mom and dad announced to the kids that they were moving into a house where each child would get his or her own room. Once again, the arguing started between Jenny and Jeff. Mom got them to measure their bedroom windows with pieces of paper, and the windows had the same area. Dad got them to measure their bedroom floors with pieces of newspaper, and the floors had the same area. Even though they had the same floor space and window size, they still argued. Jill spoke up, indicating she had the best room because it was farthest from Jenny and Jeff, and closest to Fudge, the family cat.

New York: Harper Collins Publishers, 2001   ISBN: 0-06-028918-X

Concepts or Skills

- Area

Objectives

- Determine the area of a rectangular region
- Construct a rectangle with a specified region

Materials Needed

- Unifix Cubes
- 1 cm Grid Paper, page 48
- 2 cm Grid Paper, page 49
- Manila paper

Activity 1

Have students cut out different size rectangles from manila paper. Have them place them on 1 cm grid paper, trace around the rectangle and determine the approximate area of their rectangles in square centimeters.

Activity 2

Cut out various size rectangles from 1 cm grid paper and place five in plastic bags. In cooperative groups, have students find the areas of the rectangles by counting the squares. Discuss other ways they might find the area: repeated addition, finding the product of length and width.

Activity 3

Divide the class into groups of three or four. Give each group 24 Unifix Cubes and a 2 cm grid paper. Show students that one face of a cube will fit on a square of the grid paper. The area is one square unit. Write a number of the chalkboard or overhead projector, such as 6 square units. Have the groups find different rectangles that have the specified area by placing the cubes on the grid sheet.

The table below shows the possible dimensions of rectangles up through an area of 24 square units.

AREA	DIMENSIONS
1	1x1
2	1x2, 2x1
3	1x3, 3x1
4	1x4, 4x1, 2x2
5	1x5, 5x1
6	1x6, 6x1, 2x3, 3x2
7	1x7, 7x1
8	1x8, 8x1, 2x4, 4x2
9	1x9, 9x1, 3x3
10	1x10, 10x1, 2x5, 5x2
11	1x11, 11x1
12	1x12, 12x1, 2x6, 6x2, 3x4, 4x3
13	1x13, 13x1
14	1x14, 14x1, 2x7, 7x2
15	1x15, 15x1, 3x5, 5x3
16	1x16, 16x1, 2x8, 8x2, 4x4
17	1x17, 17x1
18	1x18, 18x1, 2x9, 9x2, 3x6, 6x3
19	1x19, 19x1
20	1x20, 20x1, 2x10, 10x2, 4x5, 5x4
21	1x21, 21x1, 3x7, 7x3
22	1x22, 22x1, 2x11, 11x2
23	1x23, 23x1
24	1x24, 24x1, 2x12, 12x2, 3x8, 8x3, 4x6, 6x4

Discuss any patterns they observe. For example, a 2x3 rectangle is the same as a 3x2 rectangle.

Activity 4

For upper grades, have students measure the length and width of various rectangles and use the area formula $A = l \times w$ to find the area.

Activity 5

Have students use rulers to draw the diagonals of various rectangles. Have them find the areas of the resulting right triangles.

$$A = \tfrac{1}{2} \, l \times w$$

Writing and Communicating

Have students write about or discuss the differences between the area of a rectangle with a given length and width and the area of a right triangle with the same length and width.

Have students write about the results they discovered in Activity 3. What areas have the fewest number of rectangles that can be made? Note that for prime numbers (2, 3, 5, 7, 11, 13, 17, 19, 23) there are only two rectangles that can be constructed.

Assessment

Have students find the areas of given rectangles using Unifix Cubes as square units.

Have students find the areas of rectangles using a ruler to find the length and width, and then apply the area formula for rectangles.

Have students construct all rectangles for a given area.

Internet Links

www.mathforum.org/math.topics.html
www.gouchercenter.edu/jcampf/patterns.htm
www.mikestrong.com/doghouse/index.html

Notes:

CAPTAIN INVINCIBLE AND THE SPACE SHAPES

Story Summary

Sam is dreaming that he is Captain Invincible in his spaceship, the Hawk. As he returns to earth with his dog, Comet, Sam faces all types of perils. Using three-dimensional shapes (cube, cone, square pyramid, rectangular prism, sphere and cylinder), he is able to evade all of the dangerous objects on the return trip. He finally awakens and sees all the different shapes he used on his dream journey.

New York: Harper Collins Publishers, 2001   ISBN: 0-06-028022-0

Concepts or Skills

- Vertices, faces and edges of polyhedrons
- Three-dimensional figures: cube, rectangular prism, cone, sphere, cylinder, square pyramid

Objectives

- Identify a cube, rectangular prism, cone, sphere, cylinder and square pyramid
- Identify the polygonal faces of a cube, rectangular prism and square pyramid
- Contrast the differences between a pyramid and a cone
- Contrast the differences between a rectangular prism and a cylinder
- Contrast the differences between a circle and a sphere
- Contrast the differences between a square and a cube
- Contrast the difference between a polygon and a polyhedron
- Identify and count the vertices, edges and faces of a polyhedron

Materials Needed

- Captain Invincible Shapes Pages, pages 55-56
- Toothpicks and gumdrops, Geofix®, or straws and pipe cleaners
- Newspapers, catalogs, magazines, or grocery ads

Activity 1

Have students go on a three-dimensional geometry walk through the school building or outside on school grounds. Let students identify three-dimensional objects related to the story.

Activity 2

Have students write stories of their choice, illustrating them with shapes from their story. Share stories with each other.

Activity 3

Have students bring in different types of three-dimensional shapes from home, such as paper towels or cans (cylinders), tissue boxes or cereal boxes (rectangular prisms).

Activity 4

Make copies of the Captain Invincible Shapes pages for each student. Have them cut the shapes out, fold on the lines, and tape or glue the shapes together. Discuss the names of each shape and what polygons are used to make the shape.

Activity 5

Have students use toothpicks and gumdrops, Geofix Shapes, or straws and pipe cleaners to construct three-dimensional shapes. Discuss the number of vertices, edges and faces. Have students make and complete a chart similar to the one shown below.

Name of Shape	Vertices	Faces	Edges

Activity 6

Give students newspapers, catalogs, magazines, or grocery ads. Have them cut different photos of three-dimensional objects and categorize the shapes

Activity 7

In small groups, have students construct a spaceship model using various three-dimensional shapes. Discuss the shapes that each group used.

Writing and Communicating

Have students write about or discuss the differences between two-dimensional objects and three-dimensional objects.

Assessment

Have students name three-dimensional objects in a set.

Have students identify faces, edges and vertices for a given polyhedron.

Internet Links

www.nasa.gov

www.seds.org/nineplanets/nineplanets

marsprogram.jpl.nasa.gov

www.spaceflight.nasa.gov

www.mathforum.org/math.topics.html

Notes:

COYOTES ALL AROUND

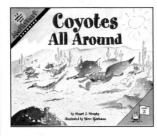

Story Summary

It is about lunch time and Clever Coyote is hungry. First, there are many roadrunners in the desert. Her friends count them, and while they are doing so, she tries to capture one. Clever Coyote misses. The other coyotes announce how many each counted. While Careful Coyote begins to add the numbers, Clever announces she can come close to the total by estimating in her head. Her estimate is two over. Next, come the lizards. She tries to capture one for lunch, but again misses. In a similar fashion, Clever estimates the number of lizards. This time, her estimate is three less than the exact number. Clumsy Coyote suggests that there are lots of grasshoppers. Clever's estimate is one more than the exact number, but she misses catching one. Finally, the coyote friends go after beetles, but they are very fast. As the friends attempt to catch some, they all end up on top of Clever Coyote. She gives up on lunch.

New York: Harper Collins Publishers, 2003   ISBN: 0-06-051531-7

Concepts or Skills

- Rounding to tens

Objectives

- Estimate
- Round numbers to the nearest ten

Materials Needed

- Unifix Cubes
- Three in a Row Grid and Cards, page 57
- Small paper plates
- 0–99 number tiles

Activity 1

Have students play the game Three in a Row, listed on the last page of the book.

Make copies of the Three in a Row game sheet. Distribute the game sheet and 10 Unifix Cubes to each student. The cubes are used as markers. Divide the class into pairs. Distribute a set of Number cards to each pair.

Have the pair place the cards face down. The first student selects three cards, rounds each number to the nearest ten, and finds the sum of the rounded numbers. The student places a cube on the square showing the sum. Cards are returned face down and mixed. The second student selects three cards and repeats the procedure. The game continues until one player gets three cubes in a row, horizontally, vertically, or diagonally.

Activity 2

The book provides several pieces of information about roadrunners, coyotes and grasshoppers. Pose problems for the students using the information.

- A roadrunner can run about 15 miles per hour. How many miles could it cover in 3 hours?
- How long would it take a roadrunner to cover 90 miles?

- Coyotes are about 4 feet long as an adult. First, have students mark off 4 feet on the floor. How many inches are in 4 feet?
- Grasshoppers can jump a distance of 200 times their body length. Have students research the length of a grasshopper's body. Then find out the length of a jump.

Activity 3

Take ten small paper plates and label them 10, 20, 30, 40, 50, 60, 70, 80, 90, 100 with a marking pen. On tongue depressors or craft sticks, write different numbers. Give each student a stick, have them round the number to the nearest ten, and place the stick on the appropriate plate. Continue daily until students understand rounding to tens.

Extend the activity to rounding to the nearest hundred.

Activity 4

As a teacher directed activity, create four two-digit numbers with number tiles. Have students round the nearest ten and find an estimate for the sum.

2	7	→ 30
8	3	→ 80
5	6	→ 60
1	9	→ 20

Estimate = 190

Divide the class into groups of two or three. Give each group a set of number tiles. Have one student in the group create the two-digit numbers with the tiles, and the other students round each number to the nearest ten and find an estimate for the sum.

For students in the upper grades, have students create one four-digit number and two three-digit numbers. Other students round to the nearest hundred and find an estimate for the sum.

Activity 5

Each week fill a large plastic container with any of the following objects:

- For younger children: rocks, large marshmallows, plastic eggs, small potatoes
- For older children: pennies, buttons, candy corn, beans, jelly beans, straws, plastic money, Unifix Cubes, different types of wrapped candy, marbles, rocks, sea shells

Have children estimate the number of objects in the jar and write the estimate on a card to submit by a specific day. At the end of the week, have students count the objects in the jar.

Discuss what influenced them to estimate as they did. Was it the size of the items? The size of the jar? The shape of the object?

After several weeks of estimating, have one student take the jar home and fill it with some object for classmates to estimate.

Activity 6

Divide students into groups of two. Give each group a grocery ad from the newspaper. Have students round prices to the nearest ten cents and estimate the cost for the objects on the page.

Writing and Communicating

Have students write about or discuss how rounding prices at the grocery store helps in determining the total bill.

> ### Assessment
> Give students ten numbers that will round to 0, 10, 20,. . .90, 100, and have them round each number.

Internet Links

www.williamcalvin.com

www.factmonster.com

Notes:

ELEVATOR MAGIC

Story Summary

Ben arrives at his mother's place of work, an office building with ten floors. As he and his mom descend in the elevator, his subtraction skills come in handy. He figures out which buttons to push on the elevator to go down. As he goes down, Ben sees some strange sites on the elevator ride. He sees chickens and a tractor in the bank, racecars in a delivery store, and a rock band playing in a candy store. Does Ben have an imagination?

New York: Harper Collins Publishers, 1997   ISBN: 0-06-026774-7

Concepts or Skills

- Basic subtraction facts

Objectives

- Write basic subtraction facts
- State and/or write ordinal numbers from first to tenth
- Draw lines of symmetry for various figures

Materials Needed

- Unifix Cubes
- Dice
- Large cups
- Two-color counters
- Crayons
- Manila paper
- Plastic coins or real money

Activity 1

Give each student ten Unifix Cubes of different colors and have them make a bar with the cubes. Everyone's bar may be different. As students hold their bars vertically on their desks, ask questions such as the following:

- What color do you have for the 1st floor? 10th floor? 6th floor?
- If you are on the 5th floor and you go up 2 floors, what floor are you on?
- If you are on the 8th floor and you go down 5 floors, what floor are you on?

Have students write addition or subtraction number sentences for each situation.

Activity 2

Play Cup the Difference! Give each pair of students one cup and some two-color counters. If you are working on the fact family for 5, have students place 5 counters in the cup. One student shakes the cup and spills the counter onto the desk. The other student writes a subtraction sentence, where red counters are subtracted from the total number of counters. For example:

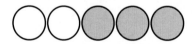

Number Sentence: 5 – 3 = 2

Now, play is reversed. The second student shakes and spills the cup, and the first student writes the corresponding number sentence. Play several times. Discuss their results in class, recording the different number sentences that students obtained. Repeat the activity with a different number of counters in the cup.

Didax® Educational Resources

Activity 3

Play Minus the Cube! Divide the class into groups of four. Give each group a regular six-sided die. Give each student ten Unifix Cubes. One student begins Round 1 and rolls the die. For the number showing on the die, the student subtracts the corresponding number of Unifix Cubes from the ten cubes and writes the subtraction number sentence. Each student in the group repeats a similar process.

Now, the game repeats. However, a student may not be able to complete a number sentence. If the student has only five cubes from Round 1 and rolls a six, he/she cannot remove 6 cubes. The subtraction number sentence that the student would write is 5 - 0 = 5.

Have students play two or three rounds. The player with the least number of cubes is the winner of the game. Have students play the game again.

For higher grades, begin with more Unifix Cubes.

Activity 4

Give students a piece of paper and have them fold it into fourths. Pose four basic subtraction equations. Have students write one of the equations at the bottom of each rectangle. Then, have students illustrate with a picture the corresponding equation.

Activity 5

Give each student ten pennies. Have them cup the pennies in their hand and drop them on their desks. Have them subtract the number of pennies showing heads from the total of ten, and then write the corresponding subtraction fact.

Writing and Communicating

Have students write a story about a trip to a tall building with an elevator.

Assessment

Give students a specific number of Unifix Cubes. Toss a die. Have students perform the subtraction and write the corresponding subtraction sentence.

Internet Link

www.sfrc.ufl.edu/Extension/pubtxt/cir1186.htm

Notes:

A FAIR BEAR SHARE

Story Summary

A mother bear wants to bake a blueberry pie for her four cubs. She sends the four cubs out to gather the ingredients: nuts, blueberries and seeds. One of the cubs always wants to play, and therefore, never collects any ingredients. As the other cubs continue to collect needed ingredients, they count their supplies by adding and regrouping. Unfortunately, since only three cubs gathered the ingredients, there wasn't enough to make the special pie. Since the fourth cub didn't help, she is sorry for playing and not working. She goes out into the forest and field and collects the additional ingredients necessary to make the pie. Mama Bear makes the pie and the cubs all get a fair bear share of the pie.

New York: Harper Collins Publishers, 1998   ISBN: 0-06-027438-7

Concepts or Skills

- Regrouping

Objectives

- Accurately estimate the number of objects in a container
- Regroup when necessary in an addition problem
- Group objects by tens

Materials Needed

- Unifix Cubes
- Base 10 blocks
- Plastic jars containing nuts or seeds

Activity 1

Display several plastic jars containing objects. Have students estimate the number of objects in each jar. Determine which student has the closest estimate by counting the objects, forming groups of 10 and regrouping when necessary.

Activity 2

Play Making Tens. Give each student a sheet of paper, along with approximately 25 base 10 unit cubes and 2 base 10 longs. Have students create a chart as shown below, labeled Tens and Ones. Divide the class into pairs of students and give a six-sided die to each pair. The game rules are:

- Each student rolls the die and the player with the lowest number goes first.
- The first player rolls the die and then puts the corresponding number of unit cubes in the Ones column.
- The second player rolls the die and also places the corresponding number of cubes in his/her column.
- Play continues in the same manner. As players get 10 unit cubes in the Ones column, they trade for one long, placing it in the Tens column.
- The player that reaches a predetermined number, such as 20, is the winner.

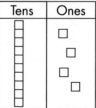

Activity 3

Play Making Tens. Give each student a Tens and Ones Chart, along with approximately 25 Unifix Cubes of one color and 2 cubes of a different color. Divide the class into pairs of students, and give a six-sided die to each pair. The game rules are:

- Each student rolls the die and the player with the lowest number goes first.
- The first player rolls the die and then puts the corresponding number of Unifix Cubes in the Ones column.
- The second player rolls the die and also places the corresponding number of cubes in his/her column.
- Play continues in the same manner. As players get 10 cubes of one color in the Ones column, they trade for one cube of the other color, placing it the Tens column.
- The player that first reaches a predetermined number such as 20 is the winner.

Notes:

Writing and Communicating

Have students write about or discuss why we need to learn to count.

Assessment

Have students model a given number with base 10 blocks.

Have students add to given numbers with or without base 10 blocks.

Internet Link

www.mathforum.org/math.topics.html

GET UP AND GO!

Story Summary

The story is about a little girl and her daily routine getting ready for school. Each event in her routine takes a certain number of minutes. Sammie, her dog, is also part of the story. As the story unfolds, the girl and dog keep track of the time by graphing on a time line that shows the total number of minutes. She finally gets out of the house in 36 minutes, and Sammie has the house the rest of the day.

New York: Harper Collins Publishers, 1996   ISBN: 0-06-025881-0

Concepts or Skills

- Timelines
- Addition

Objectives

- Construct a timeline from given data
- Interpret information from a timeline
- Construct a daily timeline

Materials Needed

- Unifix Cubes
- Bulletin board paper
- Journals
- Assorted biographies of famous people
- Internet

Activity 1

Give students sheets of bulletin board paper. Have them create a timeline of their years since they were born. Show students how to draw the year marks and highlighting special activities or events in their lives. Let students share with the class.

Activity 2

Have students read about a famous person of their choice. Have them construct a timeline of the events in the person's life. Students can use the Internet to find information about the person.

Activity 3

Have students construct a timeline of what they do each day after school. Have them compare their timelines with other students in class. How are after school hours different? Same?

Activity 4

Much like the story, have students make a list of what they do each morning before going to school. Have them estimate how much time it takes for each activity, and then determine how much time it takes to get ready for school. Make a class graph showing the number of minutes for each student.

Activity 5

Give each pair of students approximately 20 Unifix Cubes. Present a story similar to Get Up and Go! As the time for each event is read, have students model the time with Unifix Cubes, one cube for each minute. The same color of cube should be used for a specific event, then a different color for another event. Once completed, students can determine the total number of minutes.

Activity 6

Have students write an equation showing the sum of the minutes for the girl to get ready: $5 + 3 + 8 + 2 + 6 + 7 + 4 + 1 = 36$. Have students write other equations with a sum of 36. Share these equations with the class.

Writing and Communicating

Have students write in their journals what the graph says about their class getting ready for school.

Have students write about why a timeline is helpful to interpret information.

Have students write about how a timeline can help them be on time.

Have students write about what a timeline can tell me about our world.

Assessment

Have students construct and interpret a timeline for a given set of data.

Internet Link

www.biography.com

Notes:

GIVE ME HALF!

Story Summary

Brother and sister do not like to share their food or drink, but mom and dad intervene to force them to share one-half of everything. The arguments continue until a food fight erupts. Now, they must clean up the room, but realize that if they both help, along with Buddy, their dog, in will only take half the time.

New York: Harper Collins Publishers, 1996   ISBN: 0-06-446701-5

Concepts or Skills

- Half and whole
- Addition of fractions

Objectives

- Determine one-half or one-third of a set
- Model and write a number sentence
  1/2 + 1/2 = 1 or 1/3 + 1/3 + 1/3 = 1
- Shade one-half or one-third of a diagram

Materials Needed

- Animal crackers, Teddy Grahams, Goldfish Colors Crackers, or Scooby-Doo Cheddar Crackers
- Unifix Cubes
- Fraction Diagram Cards, page 58
- Show Me 1/2 Grids, page 59
- Show Me 1/3 Grids, page 60

Activity 1

Give each pair of students an even number of animal crackers. Have students separate the crackers so that each student has one-half of the crackers. Write the number sentence 1/2 + 1/2 = 1 on the board. Discuss the idea that if the two half sets are combined, the whole set is obtained.

Activity 2

Give each group of three students a number of crackers that is a multiple of three (3, 6, 9, 12, …). Have students separate the crackers so that each student has one-third of the crackers. Write the number sentence 1/3 + 1/3 + 1/3 = 1 on the board. Discuss the idea that if the three sets containing one-third of the crackers are combined, we obtain the whole set.

Activity 3

Make a transparency of the Fraction Diagram Cards. Make copies of the Fraction Diagram Cards on cardstock and cut them apart. Give each student a deck of cards, have them shuffle the cards, and then lay them face up on his/her desk.

Have students make matches of cards showing equivalent fractions (diagrams). If the fraction cards are used, then a fraction can be matched with a diagram. Examples of matches are shown below.

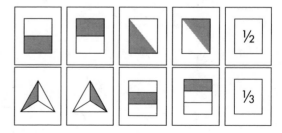

Activity 4

Make copies of the Show Me Half grids and distribute to each student. Have students use Unifix Cubes to cover one-half of each grid.

Depending upon student ability, repeat the activity using the Show Me One-Third grid.

Activity 5

Take students on a fraction walk, inside or outside of the school. Have students list things that could come in halves, thirds, fourths and so on. Keep a list as you take the walk.

Activity 6

Group students in half by one attribute, without telling how the groups were formed. Have students discuss how the groups were formed. If you have an odd number of students in class, have one student be a recorder. After discussing the fraction 1/2, have students create other ways to divide the class into halves, thirds, or even fourths. Is it always possible? Let them try making these divisions. What fraction have they made?

Writing and Communicating

Give students the prompt "If I know how that one-half of something is (a number, such as 2), then the whole is _____."

Give the students the prompt "If I know that the whole of something is (a number, such 6), then one-half of this is _____."

Repeat the previous two prompts using one-third.

Have students draw pictures or write about one-half (or one-third) of something in the classroom.

Assessment

Give children a selected set of animal crackers or Unifix Cubes and have them show one-half of the set.

Have students distinguish between one-half and one-third on various Fraction Diagram Cards.

Internet Links

www.webmath.com

www.pbs.org/teachersource/mathline/concepts/asia/activity2.shtm

Notes:

LET'S FLY A KITE

Story Summary

Bob and Hannah are brother and sister that often do not like to share things. Their babysitter, Laura, comes to the house and asks them if they would like to go to the beach to fly a kite. Of course, both siblings do. First, they must make the kite and decorate it. They learn that their kite has a line of symmetry that provides equal areas to decorate.

Even on the trip to the beach, they learn that the backseat of Laura's car has a line of symmetry that gives Bob and Hannah the same area to sit. When they reach the beach, they also find that their beach blanket and sandwich have lines of symmetry that provide each with the same area to sit and the same amount to eat.

New York: Harper Collins Publishers, 2000   ISBN: 0-06-028034-4

Concepts or Skills

- Symmetry
- Mirror image (line reflection image)

Objectives

- Describe a line of symmetry
- Find lines of symmetry for various figures
- Draw lines of symmetry for various figures

Materials Needed

- Unifix Cubes
- Manila paper
- Pattern blocks
- Crayons or colored pencils
- Construction paper
- Symmetry Grids, page 61
- Shapes Page, page 62

Activity 1

Distribute a sheet of manila paper that has a kite shape drawn on it. Have students draw a line of symmetry, and then decorate each triangular region so that one side is the mirror image of the other side.

Activity 2

Draw a line on a 8.5 x 11 inch paper. Make copies of the page and distribute to each student. Hand out pattern blocks to pairs of students. Have one student create a pattern on one side of the line. The other student constructs the mirror image of the pattern on the other side of the line. For example:

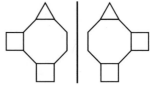

A similar activity could be done with older students using geoboards.

Activity 3

Have students go on a symmetry walk through the school. Can they see any objects that are symmetrical? Let students make a list when they return to the classroom. Share with classmates.

Activity 4

Make copies of the Shapes page on construction paper. Cut out the different shapes. Fold on the line of symmetry and cut the shapes into two parts. Place

Didax® Educational Resources

the shapes in a bag and have each student draw a shape from the bag. Once all students have a shape, let them find the matching piece.

As an individual activity, one student matches all the shapes in the bag.

Activity 5
Give each pair of students several Unifix Cubes and the Symmetry Grid Page. Similar to Activity 2, one student creates a design on one side and the second student constructs the mirror image on the other side.

Activity 6
For older students, have them find logos that have lines of symmetry. Here are some examples:

- McDonald's arch
- Mercedes-Benz automobile
- Honda automobile
- Hilton Hotel
- Chrysler automobile

Activity 7
For younger students, look at the capital letters of the alphabet. Which letters have lines of symmetry? Which have more than one?

1 line of symmetry: A B C D E M T U V W Y

2 lines of symmetry: H I X

Infinite number of lines of symmetry: O

Note, depending on how an X is made, it could have 4 lines of symmetry.

Note, depending on how a K or Q is made, each could have 1 line of symmetry.

Writing and Communicating
Have students discuss or write about why symmetry is important

Have students respond to the prompt "If We Didn't Have Symmetry in the World."

Assessment
Provide two-dimensional shapes and have students use a ruler to draw the lines of symmetry for the shapes.

Have students create a shape with pattern blocks or Unifix Cubes that has exactly one line of symmetry.

Internet Link
www.sunshine.co.nz/nz/37/themes/j_theme11.html

Notes:

MALL MANIA

Story Summary

On Mall Mania Day, the 100th person to enter the mall wins many prizes. The chess club from the elementary school is in charge of counting people entering. Brandon and his sister come to the mall to buy a birthday present for their mom, but Brandon doesn't like to shop. He stays outside with Jonathan, who is in charge of the counting. When his sister can't find a present, she comes to get Brandon, the 100th shopper at the mall.

New York: Harper Collins Publishers, 2006   ISBN: 0-06-0055777-X

Concepts or Skills

- Skip counting by 5s
- Fact strategies including doubles, doubles plus 1 and adding to 10
- Multiplace addition
- Bar graphs

Objectives

- Count by 5s to 50
- Use the fact strategy for doubles to find a sum
- Use the fact strategy for doubles plus one to find a sum
- Write addition number sentences involving double digit addends
- Construct a simple bar graph
- Use a calculator to find sums of money in cents
- Add money amounts in dollars

Materials Needed

- Small counters
- 3 x 5 cards numbered with multiples of 5 to 100
- 0–99 or 1–100 Chart, pages 50 and 51
- Chart or bulletin board paper

Activity 1

Distribute a 1–100 chart or 0–99 chart to each child. Using small counters, have children count by 5s to 100 placing a counter on each number. Discuss the pattern that appears on their charts. There will be two columns covered, one for the 5s and one for the 10s.

Activity 2

Give each child a bar of 5 Unifix Cubes of one color. Have children take turns connecting the bars of cubes on the floor, counting by 5s as they are connected.

Activity 3

Distribute a 1–100 chart or 0–99 chart to each child. Give each child several small counters. State orally different types of problems such as the following and have children cover the result with a counter.

- 1 more than 43   20 + 30
- 2 less than 37   45 + 45
- 2 tens and 3 ones   67

Activity 4

On the inside front and back of the book are pages with illustrations of a variety of items that could be purchased at a mall. Give children a list of the items with "representative" prices for each item. Have children find a group of items that will be closest to $20 without going over. Repeat the activity for $50.

Activity 5

Form groups of two to four children. Give each group a set of 3 x 5 cards numbered with multiples of 5 to 100. Deal the cards to each child. Some children may have more than others, depending on group size. The child with the 5 card starts first by saying the number name and placing it on the table. Other children follow, saying and playing the 10 card, 15 card and so on until all cards have been played.

Activity 6
Have children calculate the number of people entering the mall on each side. Then provide graph paper to make a simple bar graph showing the results. Discuss which entrance had the most/least people entering.

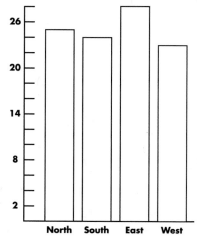

Activity 7
Display weekly vocabulary words or mathematics vocabulary words on the overhead. Distribute a sheet containing the monetary value of each letter of the alphabet as shown below. Have children find the monetary value of each word and answer the following questions.

- What is the word with the least value in cents?
- What is the word with the most value in cents?
- What word is has a value closest to $1?

Letter Values
1¢:	a	e	i	m	q	v	x	t
5¢:	c	d	f	h	j	n	p	r	u	z
10¢:	b	g	k	l	o	s	w	y

Activity 8
Take your children to the mall for Mall Math. Students should be grouped in groups of 3, with a parent volunteer with each group. Each group should have a clipboard, calculator, and paper and pencils. Questions for investigating should be on a separate sheet for each group. Have children do the following activities in the mall. You will need to let the mall representatives know that your coming and that students will not be going into stores. Children should have the following activities on sheets:

Activity I: Have children look for patterns. Have them describe what type of patterns they found. They should list where they found the pattern in the mall.

Activity II: Have children find a repeating pattern. Where did you find the pattern? List.

Activity III: If there are stairs in the mall somewhere, have children estimate how many stairs are on the staircase. Then, have the children count the stairs. Who was closest to their estimate?

Activity IV: Go to an escalator. Have children estimate how many stairs are on the escalator? How will they count the stairs to determine how close their estimate was? How do you count them if the escalator is moving? Children will need to figure this out before estimating and counting.

Activity V: In groups of two or three, have children tally the number of people that enter various stores during a 15-minute time span. Have children graph class data upon returning to class. Discuss findings.

Activity VI: Give each group of students a sheet containing the following shapes: circle, triangle, square and rectangle. Have children list places in the mall where these shapes were found.

Writing and Communicating
Have children write about mathematics in the mall.

Have children write about skip counting with a number other than 5.

Assessment

Have each child orally count by 5s to a given number.

Orally present problems to children and have them cover results on a 0–99 chart.

MORE OR LESS

Story Summary

Principal Shaw is retiring from Bayside School. The neighborhood throws a big picnic in his honor, with lots of games to play. Eddie sat in a dunk tank for the game "Let Eddie Guess Your Age!" Clara was in Eddie's class. He guessed her age in three questions…no prize! Clara tried other games, but still she did not win a prize. All the while, Eddie was guessing ages without getting dunked. Finally, Eddie met his match…he had to ask six questions, and get dunked. It was Principal Shaw. He won a prize and gave it to Clara, his granddaugher.

New York: Harper Collins Publishers, 2005   ISBN: 0-06-053165-7

Concepts or Skills

- Comparing numbers
- Even/odd numbers
- Reasoning
- Problem solving
- Estimation

Objectives

- Identify an even or odd number
- Compare two or more numbers
- Use logical reasoning
- Estimate various quantities of objects

Materials Needed

- Paper Number Lines (0–20, 0–40), page 63
- 0–99 Chart, page 50
- Clues Number Lines, page 64

Activity 1

Cut out the 0–20 or 0–40 number lines. Tape the pieces together to make the lines. Distribute a number line to each child.

Play "Guess My Number!" Write a number on paper. Call on children to ask questions about your number. Tally the number of questions they ask before finally guessing your number.

In order that children understand the game, have one child write the number and you ask questions to guess the number. Use a 0–99 chart for a more difficult game of "Guess My Number!"

Activity 2

Depending on class size, write large numbers on blank paper up to half the size of the class. If there are 30 children in class, write numbers from 1 to 15.

Distribute the numbers to children. Have them form a human number line in front of the rest of the class. Prepare sheets with phrases such as the following:

- Greater than 6
- Less than 12
- Greater than 3 and less than 7
- Greater than 13 and less than 15
- 9

Show the human number line on one of the sheets without children at their seats seeing the phrase. If a child's number satisfies the phrase, the child steps forward. Children at their seats attempt to guess the phrase.

As mathematical symbols and variables are introduced in higher grades, write phrases such as the following:

- $N = 7$
- $N > 10$
- $N < 5$
- $4 < N < 8$
- $N \neq 3$

Activity 3

Play the game of "Clues." Distribute a Clues Number Line to each child.

To provide initial instructions for children, have them make a number line from 1 to 20 on a piece of paper. Draw a similar number line on the chalkboard or overhead. Provide a set of clues, one at a time, and show children how to mark their number lines to determine the number. Here is a set of clues to use:

- The number is greater than 10. (Cross out 1 through 10.)
- The number is less than 15. (Cross out 15, 16, 17, 18, 19 and 20.)
- The number is odd. (Cross out 12 and 14.)
- The number is the first remaining number. (11)

Write out clues for specific numbers as suggested at the end of the story. Use the range of numbers marked on the "Clues" number lines. Read the clues for each number. Provide time for children to mark their number lines.

Activity 4

To begin, write a number on the chalkboard. Place several overhead counters on the projector. Turn the projector on and then off, giving children just a second or two to view the overhead. Ask children if the number of counters was more or less than the given amount you stated. Turn the overhead back on for a longer period of time, then turn it off. Ask the same question. Repeat the activity for several different sets of counters.

Activity 5

Unwrap a package (500 sheets) of copy paper. Take a part of the package with more than 200 pages and ask children if there are more or less than 200 sheets of paper. Have one or two children count out 100 sheets of paper and hold up their combined pile against your original stack. Children will see that there are more than 200 sheets.

Writing and Communicating

Have children write about other words that are synonymous with greater than, less than, and equal to. Make a word wall with their findings.

Assessment

Give each child a set of clues and a number line. Have the child find the number determined by the clues.

Give each child a number line. Orally present a number phrase such as "The numbers less than 10." With or without the number line, have the child write or state the resulting numbers.

Internet Link

www.mathforum.org/funpow/

Notes:

PEPPER'S JOURNAL

Story Summary

This book provides an excellent introduction to journal writing. Joey and Lisa eagerly await their grandmother's cat, Snowy, to have her kittens. Finally, on March 6, Snowy has three kittens. Although the children can't hold the kittens for one month, they do get to visit and touch them. When one month has passed, Lisa and Joey select the kitten they want to keep. Since he is black and white, they decide to name him Pepper. As the story continues, Lisa adds to her journal, writing entries each month until Pepper celebrates his one year birthday.

New York: Harper Collins Publishers, 2000   ISBN: 0-06-027618-5

Concepts or Skills

- Time
- Timelines
- Calendars
- Graphing
- Probability

Objectives

- Name the months of the year
- Describe a given timeline
- Construct a timeline
- Construct an object or picture graph

Materials Needed

- String
- Blank cards
- Plastic coins
- One-month calendar
- Plastic bags

Activity 1

Introduce the idea of a journal to children. Ask them if they have ever kept a journal. Discuss famous people who have kept daily journals about their lives (presidents and first ladies, scientists, astronauts). Discuss how a journal serves as a timeline.

Activity 2

Have children construct their own timeline, from birth to the current year. Have them illustrate their timelines by drawing or using actual photographs. Share the timelines in class.

Activity 3

Have students make a one-week calendar. Have them write or illustrate certain events that occurred each day.

Activity 4

For older students, make a one-month calendar and record daily activities. After the month is complete, discuss or have children write about the following:

- How much time did I spend watching TV?
- How much time did I spend reading?
- How much time did I spend doing chores?
- How much time did I spend doing homework?

Activity 5

In class, have students make a people graph for the month, year, or day when they were born. Have them interpret the graph and write in their journals.

Activity 6

Using a one-month calendar and small plastic bags for each day, place pennies in the bag corresponding to the date. How much money will we have at the end of the month? For months with 30 days, we will have $4.65; for 31 days, $4.96; for 28 days, $4.06; and for 29 days, $4.35.

Pose questions such as the following:

- How much money will we have in one year? Two years?
- When will we have $100 or more?

Prepare a letter to parents related to this activity asking for help in finding the monetary value of their children's birth date based on the above situation. For example, if the birth date is 14, then the amount of money would be $1 + 2 + 3 \dots + 14 = 105$¢ $= \$1.05$.

Activity 7

The story provides some information about left-pawed and right-pawed cats.

For students that have cats as pets, have students hang a pet toy on a string and observe which paw the cat typically uses to play with the toy. Record the data in class and compare the results to the information presented: $4/10 = 2/5$ right-pawed, $4/10 = 2/5$ left-pawed; $2/10 = 1/5$ no difference.

Writing and Communicating

Have students write about:

- the importance of a calendar.
- the origin of a calendar.
- their favorite month.
- naming a month.

Assessment

Provide a timeline for some particular event and have students interpret the timeline.

Write the names of the months or the names of the days of the week on cards and have students place them in the correct order.

Internet Link

www.healthypet.com

Notes:

PROBABLY PISTACHIO

Story Summary

It's Monday, the first day of the school week, and nothing is going right for Jack. Dad fixes lunches for the kids, but they never know what he will fix. Jack loves pastrami, and he knows his friend Emma had pastrami every day, except Thursday. Since it was Monday, he trades lunches with Emma, only to find out that her dad has fixed a liverwurst sandwich, which he hates.

After school, Jack and his friend Alex go to soccer practice. The coach always has the boys count off by twos. Jack and Alex line up with one boy between them so they will end up on the same team. Today, however, the coach has them count off by threes, and they end up on different teams. When practice is over, their coach gives them a snack. Jack loves popcorn, but when he reaches into the basket, he grabs a bag of pretzels, one of his least favorite foods. What else can go wrong?

When Jack enters his home, he smells something like pizza, one of his favorite foods, but dad has fixed spaghetti and meatballs. Soon, his mom comes home and says she has a surprise for the family. She pulls out a tub of chocolate ice cream, his sister's favorite flavor but not his. But his mom pulls out a second tub, and it is his favorite ice cream...pistachio.

New York: Harper Collins Publishers, 2001   ISBN: 0-06-028028-X

Concepts or Skills

- Probability
- Predictions
- Equally likely events

Objectives

- Determine the probability of a simple event
- Make predictions based on previous events
- Collect and analyze data through a survey

Materials Needed

- Unifix Cubes
- Matchbox cars/trucks
- Measuring tape or meter stick

Activity 1

In the story, Jack daydreams about Emma's usual lunches. On Monday, Tuesday, Wednesday and Friday, she has pastrami sandwiches, while on Thursday, she has soup. Have each student write "pastrami" on four cards and "soup" on one card. Fold the cards and have students place the cards in a paper bag. On the board or overhead, write the five school days, Monday through Friday. One at a time, have each student draw a card from the bag and record on the board for each day whether Jack would have a sandwich or soup. How many times did Jack get a pastrami sandwich on Thursday? How many times did he have soup?

Have students make predications: If we repeated the activity, what do you think the results would look like? Repeat the activity.

Activity 2

Give each student three Unifix Cubes of different colors. All students should have the same three colors. Have them place the cubes in a line. On the board, have students record the order of their cubes. Discuss the results shown in the table.

For three different colors, there are six different ways to arrange them:

The activity can be extended to four Unifix Cubes of different colors. In this case, there are 24 different arrangements of the cubes.

Activity 3

For practice on making predictions, play What's My Price? Display cans or bags of food, with prices marked on the bottom. The number of cans should be equal to one-half the number of students in class. Divide the class into two teams. Each team member will guess the price of an item shown by the teacher. The team member whose guess is closest to the actual price will receive a point. Continue playing the game until all items have been priced. The team with the most points wins the game.

Activity 4

Have students bring a Matchbox car or truck. Encourage them to bring extras for those students that do not have one. On a smooth surface, place a piece of masking tape, marked START. Have each student practice race one or two times while the class watches. Have students make predictions as to whose car/truck will go the longest distance. Record the predictions on the board. Before racing, ask students "Do all cars have an equal chance of winning? Why or why not?" Discuss their responses.

Now, the race begins. Have each student race his/ her car. Measure the distance from the START line to where the car/truck stopped. If the car hits the wall, or turns over, the car is out of the race. When all students have raced, take the five cars with the longest distances and race once more to determine the winner. Discuss the predictions. How many predicted correctly? How many predicted one of the top five?

Activity 5

In the story, Jack mentions his favorite sandwich, his favorite ice cream and his favorite snack. Have students collect class data for these favorites and make bar graphs.

Have students make predictions based on their graphs. If we go to a different class and ask them what their favorite ice cream flavor is, what do you think it might be? Will their answers be the same as ours?

Activity 6

At home, have students take a survey of ten (twenty) people's favorite fast food restaurant and make a bar graph showing the results. Based on the survey results, have each student predict a classmate's survey results. Record how many predictions were correct and incorrect. Discuss the results.

Activity 7

Discuss the concept of probability. Using Unifix Cubes, begin with a simple experiment, one red cube. If the cube was the only cube in a bag and one cube is drawn from the bag, what is the probability that it is red? $P(Red) = 1/1 = 1$.

Continue increasing the number of cubes and colors. If there were two cubes, one red and one blue, in a bag, and one cube is drawn, what is the probability that it is red? Blue? $P(Red) = P(Blue) = 1/2$. What is the probability that the cube is green? Since there were no green cubes in the bag, the probability is 0. $P(Green) = 0$.

Writing and Communicating

Have students discuss or write about how we can use predictions in our everyday living.

Have students conduct a survey, record the results and write about their findings.

Assessment

Show students various sets of Unifix Cubes and have them make predictions as to what color of cube might be drawn. Have students determine the probability of drawing each color from a bag.

Show students a line graph for some data and have them interpret the results in the graph. Have them predict the outcome if the experiment was repeated.

RACING AROUND

Story Summary

Mike and Marissa are entering a 15 km bike race on Perimeter Path. Their younger brother Mike wants to enter the race also, even though he has never ridden his bike that distance. He starts practicing by riding around the rectangular athletic field, and then the pentagonal zoo. Neither of these rides are close to the distance for the race. Mike gets his dad to sign the permission slip for the race. When race day comes, Mike is at the back of the group when they start racing. He hits a rock and falls. Bingo, his dog, is chasing after him. Although he wants to quit, he decides he must finish the race with Bingo. Mike finally makes the finish line, with his brother and sister cheering him on.

New York: Harper Collins Publishers, 2002   ISBN: 0-06-028913-9

Concepts or Skills

- Perimeter

Objectives

- Determine the perimeter of a polygon
- Construct a rectangle with a given perimeter

Materials Needed

- Unifix Cubes
- 2 cm Grid Paper, page 49
- Manila paper

Activity 1

Have students use the side of a Unifix Cube as a standard unit of measure to find the perimeter of their desks. Discuss the results. Do all students get the same perimeter? Why or why not?

Have them find the perimeter of other regions.

Activity 2

Give each student 6 Unifix Cubes and 2 cm graph paper. Have students arrange the cubes to make different rectangles, and then color each one. Now, have them find the perimeter of each rectangle. With 6 cubes, there are only two distinct rectangles that can be made:

- 2 x 3 (or 3 x 2) with a perimeter of 10 units
- 1 x 6 (or 6 x 1) with a perimeter of 14 units

Have students use only 4 cubes to make different rectangles and then color each one. Have them find the perimeter of each rectangle. With 4 cubes, there are also only two distinct rectangles that can be made:

- 2 x 2 with a perimeter of 8 units
- 1 x 4 (or 4 x 1) with a perimeter of 10 units

Have students explore perimeters with different numbers of cubes. Have students note for rectangles there are two pairs of sides that have the same measure. With this idea, a formula for the perimeter of a rectangle can be developed.

$$P = 2l + 2w = 2(l + w)$$

For older students, this activity also leads naturally to the discovery of prime numbers. In this case, prime numbers will be numbers for which only one distinct rectangle can be constructed. For example, using 2, 3, 5, or 7 cubes yields one distinct rectangle for each.

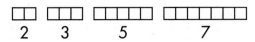

Activity 3

Using a computer graphing program, make various polygons where the length of a side is a multiple of 2 cm, the dimension of a Unifix Cube. Distribute the pages to students and have them find the perimeter of each polygon. Discuss the results.

For a regular polygon, the perimeter can be found by multiplying the length of a side times the number of sides. A square is the basic example: $P = 4 \times$ length of side.

Activity 4

Using the pages described in Activity 3 or new pages, have students use rulers to find the perimeters of polygons to the nearest centimeter.

Activity 5

Have a Measurement Day in the classroom or in the gym. Mark various items to be measured in Unifix Cubes. Give each student or group of students a record sheet. Have them estimate the dimensions of the item and then measure it. When all students have finished, discuss various perimeters. Did all students get the same perimeter? Why or why not?

Writing and Communicating

Have students write about or discuss how perimeter helps us to understand different shapes.

Have students write or discuss how perimeter is important to a house builder.

Assessment

Have students construct rectangles with a given number of Unifix Cubes, and then have them find the perimeter of each.

Give students a perimeter and have them construct a rectangle that has the specified perimeter.

Have students find the perimeter of given polygons, using rulers as measuring tools.

Internet Link

www.webmath.com

Notes:

SAME OLD HORSE

Story Summary

Hankie is a very predictable horse. He sneezes every 20 minutes, comes out of his barn on regular intervals, eats, drinks and rolls in the grass at the same times every day. His friends tease him about his predictability, so Hankie tries to change his patterns. His attempts are to no avail since his sneezes always seem to cause his daily routines to be at the same times.

New York: Harper Collins Publishers, 2005   ISBN: 0-06-055770-2

Concepts or Skills

- Making predictions
- Patterning
- Probability

Objectives

- Make a prediction based on given information
- Determine and extend a pattern

Materials Needed

- Number cubes

Activity 1

Have children do research about weather in their home city or town for the last 5 years.

- Dates for the first snow of the season or first frost.
- Total amount of snow for a given month.
- Total number of inches of rain for a given month.
- Average temperature for a given month.
- Highest/lowest temperature for a given month.

After finding information and discussing it in class, give children 3 x 5 cards, one for each prediction to be made. Have children make their prediction for each question. Children should write their names and the date on the card. At the end of the time period, check the predictions. Who was closest to the exact measure?

Activity 2

Have children predict on what day their birthday will fall in a given year, using a current calendar and figuring what day their birthday will fall on next year, in two years, in 10 years. You may want to discuss patterns as to how the prediction for a future year works.

Activity 3

Conduct a "lottery" in class. Have children write four numbers between 1 and 50 inclusive, with no repeated numbers. Write the numbers 1 through 50 on pieces of paper and place them in a paper bag. Have a child draw four numbers from the bag and record them on the chalkboard. Have children check their numbers against those drawn. How many children matched all four numbers? Three numbers? Two numbers? One number? No numbers? Discuss the class findings.

Activity 4

Distribute a number cube to each child. Children will toss the cube 24 times. Have children predict what number will appear most often. Note that mathematically each number is equally likely to appear. As children toss the cube, have them make a tally for each result. When children have finished, make a class bar graph showing the combined results. Discuss the findings. How many had correct predictions?

Activity 5

Distribute a pair of number cubes to each child. Children will toss the cubes 20 times and find the sum of each toss. First discuss the possible sums, 2 through 12. Have children predict which sum will occur most (least) often. As children toss the cubes, have them make a tally of the sum for each toss. When children have finished, make a class graph showing the combined results. Discuss the findings. How many had correct predictions? Mathematically, a sum of 7 is most likely to occur and a sum of 2 or 12 is least likely to occur.

Activity 6

On a piece of paper, have children make predictions about classmates for items such as the following:

- How many have the same birthdate?
- How many were born in the same state, city, or hospital?
- How many may have 1 sibling? 2? 3? 4? More that 4?

Writing and Communicating

Have children write about predictability in the classroom or at home.

Have children write about predictability in weather or in sports.

Assessment

Give each child a number or color pattern. Have them predict what numbers or colors will appear next.

Internet Link

www.coolmath.com

Notes:

SPUNKY MONKEYS ON PARADE

Story Summary

The Monkey Day Parade is taking place. In the parade are the marshal, majorette, 20 cycling monkeys, 30 tumbling monkeys, a 40-member monkey band, and the monkey king and queen throwing bananas to the crowd. The monkeys are riding in groups of 2, the tumbling monkeys are in groups of 3, and the monkey band is marching in groups of 4.

New York: Harper Collins Publishers, 1999   ISBN: 0-06-028024-X

Concepts or Skills

- Counting
- Counting by 2s, 3s, 4s and 5s

Objectives

- Count by 2s to a given number
- Count by 3s to a given number
- Count by 4s to a given number
- Count by 5s to a given number

Materials Needed

- Unifix Cubes
- Small counters
- 1–100 or 0–99 Chart, pages 50-51
- Large cut-out numbers for 2, 3 and 4

Activity 1

Make large cut-out numbers for 2, 3 and 4. Post these numbers in the classroom. On note cards, have students write things that come packaged in groups of 2s, 3s, or 4s. Here are some examples:

- 2s: Paper towels, Almond Joy candy, Reese's Cups, buttertubs, batteries, soda pop (3 groups of 2), one dozen eggs (6 groups of 2)
- 3s: soda pop (2 groups of 3)
- 4s: Toilet tissues, light bulbs, butter quarters, batteries

Have students post their cards under the correct number. Discuss their findings.

Activity 2

Give each student 12 Unifix Cubes. Have students count the cubes by 2s. Have them count the cubes by 3s. Have them count the cubes by 4s.

Give 24 Unifix Cubes to a group of three students. Have one student count the cubes by 2s, one by 3s and one by 4s.

Activity 3

Distribute calculators to each student. Have them skip count by 2s, 3s, 4s and 5s using the calculator. Show how to use the automatic constant feature on most basic calculators. Typically, the following chains of keys will produce multiples:

   2 + = = = = = = =

This gives 2 4 6 8 10 12 14

   3 + = = = = = = =

This gives 3 6 9 12 15 18 21

   4 + = = = = = = =

This gives 4 8 12 16 20 24 28

5 + = = = = = = =

This gives 5 10 15 20 25 30 35

As students do this, have them say each number aloud. After this activity, have children skip count by the same number again, but this time writing the numbers on paper as they count.

Activity 4
Have students determine the total number of monkeys in the parade:

   1 + 1 + 20 + 30 + 40 + 1 + 1 = 94 monkeys

Activity 5
Distribute small counters and a 1–100 or 0–99 chart to each student. Have students place a counter on 2 and skip count to 20 or higher. Discuss the patterns they observe on the charts. There are alternating columns, with even numbers covered and odd numbers uncovered. All the numbers in a column that is covered end in the same digit, either 0, 2, 4, 6, or 8. All the numbers in a column that is not covered end in the same digit, either 1, 3, 5, 7, or 9.

Have students place a counter on 3 and skip count to 30 or higher. Discuss the patterns they see on the charts. There are diagonals from the upper right of the chart to the lower left. The numbers covered are even or odd. They are the multiples of 3.

Have students place a counter on 4 and skip count to 40 or higher. Discuss the patterns they see on the charts. Again, there are diagonals running from the upper right to the lower left. The numbers covered are all even. They are the multiples of 4.

Have students place a counter on 5 and skip count to 50 or higher. Discuss the patterns they see on the charts. Two columns appear, one with numbers ending in 0 and one with numbers ending in 5.

Activity 6
Have students place a counter on 2 and skip count to 30 or higher. On the same chart, have students place a counter on 3 and skip count to 30 or higher. Discuss the numbers that have a stack of two counters. These are the common multiples of 2 and 3. All of these multiples are even. They are the multiples of 6.

Activity 7
In a small group setting, have students model the monkeys in the parade using Unifix Cubes. They should show 10 rows of 2 for the cycling monkeys, 10 rows of 3 for the tumbling monkeys, and 10 rows of 4 for the band.

Writing and Communicating
Have students find items at home or in the grocery store that are packed by 2s, 3s, 4s, 5s, or other numbers. Then have them write about the items found.

Have children write about how a calculator can help to skip count by other numbers.

Have students write a story about another parade, which includes marching with groups of 5.

Have students write about why counting is important.

Assessment
Have students skip count by 2s, 3s, 4s, or 5s to a given number.

Present students with various arrays and have them count the items.

Internet Link
www.wonlinskyweb.net/measure.htm

Notes:

THE SUNDAE SCOOP

Story Summary

At the end of the school year, the cafeteria lady, Winnie, decides to have an ice cream booth at the picnic. Lauren, James and Emily were on hand to help Winnie. The kids decided to have ice cream sundaes at their booth. Deciding which toppings, sauces and ice cream to have becomes a mathematical problem for them. Winnie lists different combinations that the kids can choose from with the toppings, sauces and ice cream they have chosen. The ice cream booth is a success, except James, Lauren and Emily didn't have their favorite toppings to make their own ice cream sundaes.

New York: Harper Collins Publishers, 2003   ISBN: 0-06-028924-4

Concepts or Skills

- Combinations
- Permutations

Objectives

- Determine the number of combinations for a given situation
- Determine the number of permutations for a given situation

Materials Needed

- Construction paper (tan, brown, pink, white)
- Crayons or colored pencils
- Manila paper

Activity 1

Have children cut circles from brown, white and pink construction paper to represent scoops of chocolate, vanilla and strawberry ice cream. Also have them cut triangles from tan construction paper to represent ice cream cones. Each child will need a cone and one each of the three flavors of ice cream.

Ask this question: How many different two-dip combinations can you make using vanilla (V), chocolate (C) and strawberry (S) ice cream?

If order does not matter (combinations), then there are only three possible arrangements on the cone:

- V-C (same as C-V) •   V-S (same as S-V)
- C-S (same as S-C)

If order does matter, then there are six possible arrangements: V-C, C-V, C-S, S-C, V-S, S-V. If the same flavor can be used twice, then there are three additional arrangements for a total of nine: V-V, C-C, S-S.

Activity 2

List on the chalkboard or overhead projector three different toppings such as nuts (N), cherries (C), or whipped cream (W). Have children illustrate each of the toppings on manila paper and cut them apart. There are several questions to pose:

A. How many possible combinations are there for one scoop of ice cream and one topping?

- V-C, V-N, V-W   • C-C, C-N, C-W
- S-C, S-C, S-W

Didax® Educational Resources

Mathematically, 3 flavors of ice cream x 3 toppings = 9 combinations. If there were two types of cones, then the number of combinations would be 2 x 3 x 3 = 18.

B. How many possible combinations are there for two scoops of ice cream and one topping?

There were three combinations with two scoops from Activity 1 and each of those could have three different toppings. There would be nine possible combinations.

C. How many possible combinations are there for three scoops of ice cream and one topping?

Since there is only one combination with three scoops (order does not matter!), there would be three combinations. Using the circles, cones, and topping illustrations, have children create the various combinations.

The same activity could be done with the following food items:

- hamburgers–pickles, mustard, ketchup
- hot dogs–mustard, relish, coney sauce
- pizza–sausage, pepperoni, onions

Activity 3

Divide the class into small groups. Distribute four Unifix Cubes of different colors to each student. Have each group explore different combinations of two cubes, then three cubes. For four cubes with groups (order does not matter) of two colors, there are six combinations.

For four cubes with groups (order does not matter) of three colors, there are four combinations.

The mathematics involved in this story is powerful, and involves topics now taught in high school. For elementary grade teachers, however, looking at patterns is the main focus of this book.

The simplest ideas come from using one cube of each color. Illustrations where order does matter are provided below.

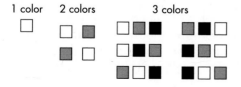

Number of Cubes	Number of Arrangements	Pattern
1	1	1
2	2	2 x 1
3	6	3 x 2 x 1
4	24	4 x 3 x 2 x 1
5	120	5 x 4 x 3 x 2 x 1

For upper grades, we can generalize the results from the table. If there are n cubes of different colors, then there are

n x (n - 1) x (n - 2) x (n - 3) x (n - 4) x ... x 2 x 1

possible arrangements of the cubes.

Mathematically, we have a shorthand way of writing this result:

n! = n x (n - 1) x (n - 2) x (n - 3) x (n - 4) x ... x 2 x 1

We read this as "n-factorial." For example, 3! = 6, 4! = 24, and 6! = 6x5x4x3x2x1 = 720.

Activity 4

Have children research when ice cream was invented. Children can also research these questions:

1. Which state eats the most ice cream? (Utah)
2. Which is the most popular flavor? (vanilla, and chocolate is the second)
3. What are the 5 top ice cream-eating countries of the world?
4. What are the most popular ice cream toppings?
5. What ice cream flavors flopped?

Activity 5

Have children research different ice cream recipes and make a class booklet.

Writing and Communicating

Have write about:

My Favorite Ice Cream Flavor

If I could invent a new flavor of ice cream, it would be…

Arranging My (#) Trucks, Dolls, or other Toys

Assessment

Present students with a set of five Unifix Cubes of different colors. Have them make groups of 3 cubes (there are 10 possible groups of 3 cubes). Have them make groups of 2 cubes (there are also 10 possible groups of 2 cubes).

Internet Link

www.ice-cream.org

SUPER SAND CASTLE SATURDAY

Story Summary

Three friends, Sarah, Juan and Laura, participate in a sand castle building contest, each using his/her own non-standard unit of measure (shovel, spoon, foot). Larry, the lifeguard, is giving prizes for the tallest tower, deepest moat and longest castle wall. As the three friends build and measure, they fail to realize that their measuring tools are not the same. When Larry blows his whistle to end the competition, he measures, their towers, moats and walls using a standard unit of measure, the inch. Now, the friends understand that everyone's shovel, spoon and foot are not the same.

New York: Harper Collins Publishers, 1999   ISBN: 0-06-027612-6

Concepts or Skills
- Measure with non-standard and standard units

Objectives
- Measure an object with non-standard units
- Measure an object with standard units
- Describe the difference between standard and non-standard units

Materials Needed
- Unifix Cubes
- Toothpicks
- Straws

Activity 1

Have students take a non-standard measuring tool of their choice and measure five different pre-selected items in the classroom. When all students have completed the task, record measurements for each of the items and discuss the differences in the results. Have students measure each item with a ruler or tape measure. Pose the question: How were the two measurements different?

Activity 2

Distribute 10 Unifix Cubes to each student. Have students measure different items on the playground. Discuss the results.

Activity 3

Give each student a toothpick. Have students measure and record the length and width of their desks, books, or other items using the toothpick. Then, have different students measure some other student's items. Discuss whether the results were the same. Discuss why it is often difficult to measure with a non-standard measuring tool.

Activity 4

Have students measure each other's height using non-standard units. Then, follow up the activity by measuring in inches.

Activity 5

Have students estimate the lengths of their feet in Unifix Cubes. After they have recorded their estimates, let them measure their feet in cubes. Ask questions: How many hit their estimate? One off?

Two off? More than two off? Estimate other lengths and repeat the questions.

Writing and Communicating

Have students discuss or write about why a system with standard units of measure is very helpful.

For older students, have them search for information about how an inch or foot was originally defined (inch: 3 barley corns, end to end; foot: the king's foot).

Assessment

Have students measure given line segments with standard and non-standard measuring tools.

Have students estimate various lengths and then measure them.

Notes:

TALLEY O'MALLEY

Story Summary

The O'Malley family, with mom, dad, Eric, Bridget and Nell, is headed to the beach for a vacation. In a typical scenario, the children become restless during the trip. Mom has the children play a tallying game with each using a favorite color, first counting cars, then t-shirts, and finally, train cars. Nell, the youngest, picks red each time. When the first two games end and the children count their tally marks, Nell loses. Finally, she wins the last game, since it happened to be the Red Line and she had the most tally marks. In her glee, Nell now wants to be called Tally O'Malley.

New York: Harper Collins Publishers, 2004   ISBN: 0-06-053162-2

Concepts or Skills

- Data collection
- Tallying
- Graphing
- Making tables
- Mode

Objectives

- Determine the mode for a set of data
- Construct a table showing a set of data
- Construct a horizontal or vertical bar graph from data collected by tallying
- Skip count by 5s

Materials Needed

- Unifix Cubes
- 2 cm Grid Paper, page 49
- Small counters
- 0–99 or 1–100 Chart, pages 50-51
- Pair of 6-sided number cubes

Activity 1

The last two pages of the book show a picture gallery of the O'Malley vacation, along with tallying for mom's coffee, dad's donuts, say cheese pictures and Nell's count of red train cars. Have students determine how many tally marks are present, orally counting by fives.

Prior to these last pages are two other pages, filled with many groups of five tally marks. Each page has 22 groups of 5 marks. For higher grades, skip count to 110.

Activity 2

Distribute counters and a 1–100 or 0–99 chart to each student. Have students place a counter on 5 and skip count to 100 (95). Discuss the patterns they observe on the charts. Two columns appear, one with numbers ending in 0 and one with numbers ending in 5.

Activity 3

Give students a prompt, such as "My favorite color is ____." Have students respond to the prompt, and then have selected students create a tallying table for the entire class on the chalkboard. Once completed, discuss the results shown in table. Discuss which color(s) appear most. Statistically, the score or number or color, in this case that appears most frequently is called the mode.

Have students create a horizontal or vertical bar graph on 2m Grid Paper with their data, using Unifix Cubes to make each bar. Discuss how to determine the mode for each graph. In a vertical bar graph, the mode corresponds to the color or number of the bar that is the tallest. In a horizontal bar graph, it corresponds to the color or number of the bar that is longest. In some cases, there may be more than one mode.

Below are examples of other prompts for tallying:

My favorite food is _____.

The month of my birthday is _____.

The number of children in my family is _____.

My pet is a(n) _____.

My favorite baseball team is the _____.

Activity 4

Give students a piece of paper and have them fold it so there are six sections. Have them mark the sections 1 through 6. Give each student a 6-sided number cube and have them roll it 30 times. Each time, they place a tally in the section corresponding to the number rolled on the number cube. Discuss the results. On the chalkboard or overhead, compile the data for all students. Discuss the idea that each number on the number cube is equally likely to appear. With 30 tosses, we would expect five tallies in each section.

For the entire class, 1/6 of the tallies should be in each section.

1 ////	2 //// //	3 ///
4 ////	5 //	6 //

As an extension, have students toss a pair of number cubes 36 times. Mark off 11 sections on a piece of paper and number the sections 2 through 12. For each toss, have students find the sum of the two numbers and place a tally in the corresponding section of the paper. Discuss individual results.

Mathematically, when a pair of number cubes are tossed there are 36 possible outcomes. The sums of the numbers in those outcomes range from 2 to 12. The resulting probabilities are shown below.

Sum	Probability		
2	1/36	7	6/36
3	2/36	8	5/36
4	3/36	9	4/36
5	4/36	10	3/36
6	5/36	11	2/36
		12	1/36

If a pair of number cubes was tossed 36 times, we would expect 1 outcome for a sum of 2, 2 outcomes

for a sum of 3, and so on. For class data, we would expect 1/36 of the tosses to be in the 2 column.

Activity 5

Have students make class tally graphs. For every graph, give each student a toothpick and have him or her glue it vertically on the appropriate section, crossing it diagonally for the fifth tally. Discuss each tally graph. Here are some titles for tally graphs:

- Favorite subject
- Favorite color
- Jobs at home
- Favorite pizza topping
- Favorite food
- Favorite holiday
- Favorite cookie
- Favorite season

Activity 6

Have students make two-row tally graphs, with headings such as the following:

- My Street Number Is: Odd, Even
- I Write with My: Left Hand, Right Hand
- My Favorite Milk Is: White Milk, Chocolate Milk
- I Was Born in: This State, Another State

My Street Number is	
Odd	//// //// ///
Even	//// //// //// //

Writing and Communicating

Have each student create his/her own survey to collect data. Once collected, each student should make a tally sheet and a bar graph. Upper grade students can write simple sentences describing the results. Here are some suggestions for surveys:

- Have you ridden in an airplane?
- Have you ever stayed in a hospital overnight?
- Have you ever taken a trip on a train?
- Do you like spinach?
- Have you ever won something?
- Have you met a famous person?
- Have you been part of a parade?
- Do you like chocolate?

Assessment

Present students with a table with tally marks. Have them determine the number of tallies for each category in the table.

Present students with a horizontal or vertical bar graph. Have them determine the mode.

Internet Link

www.gouchercenter.edu/jcampf/graphs.htm

1 CM GRID PAPER

2 CM GRID PAPER

0-99 CHART

0	1	2	3	4	5	6	7	8	9
10	11	12	13	14	15	16	17	18	19
20	21	22	23	24	25	26	27	28	29
30	31	32	33	34	35	36	37	38	39
40	41	42	43	44	45	46	47	48	49
50	51	52	53	54	55	56	57	58	59
60	61	62	63	64	65	66	67	68	69
70	71	72	73	74	75	76	77	78	79
80	81	82	83	84	85	86	87	88	89
90	91	92	93	94	95	96	97	98	99

Didax® Educational Resources

1-100 CHART

1	2	3	4	5	6	7	8	9	10
11	12	13	14	15	16	17	18	19	20
21	22	23	24	25	26	27	28	29	30
31	32	33	34	35	36	37	38	39	40
41	42	43	44	45	46	47	48	49	50
51	52	53	54	55	56	57	58	59	60
61	62	63	64	65	66	67	68	69	70
71	72	73	74	75	76	77	78	79	80
81	82	83	84	85	86	87	88	89	90
91	92	93	94	95	96	97	98	99	100

100 SQUARE GRID

TOTALING 10 CARDS

5 2	1 2	5 0	3 2	6 2
1 1	4 2	2 2	0 0	6 4
5 3	6 3	3 3	4 0	3 0
6 1	1 3	6 0	2 0	1 0
5 1				

PIZZA TOPPINGS CIRCLE

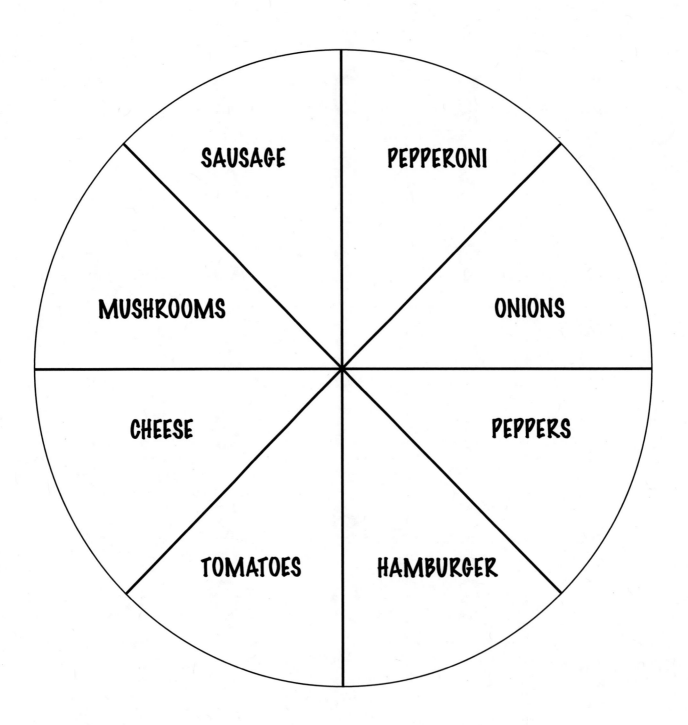

CAPTAIN INVINCIBLE SHAPES

cube

rectangular prism

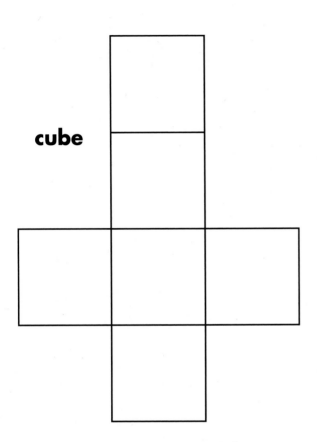

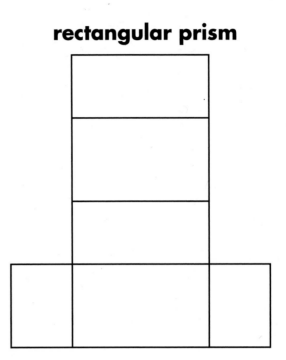

square pyramid

cone

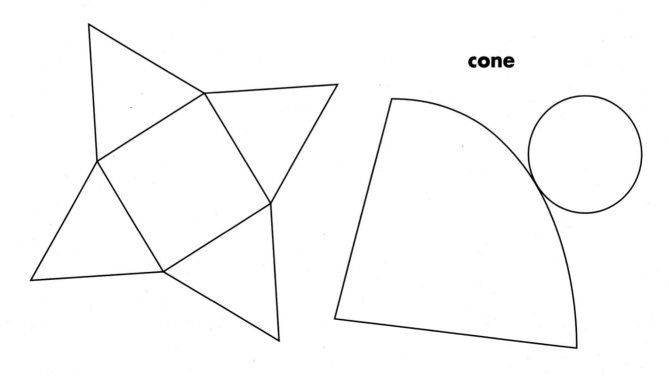

CAPTAIN INVINCIBLE SHAPES

triangular pyramid

triangular prism

cylinder

Didax® Educational Resources

3 IN A ROW GRID & CARDS

70	60	30	40
50	10	10	80
100	80	40	20
30	50	70	60

17	24	1	12	8
31	29	36	4	

FRACTION DIAGRAM CARDS

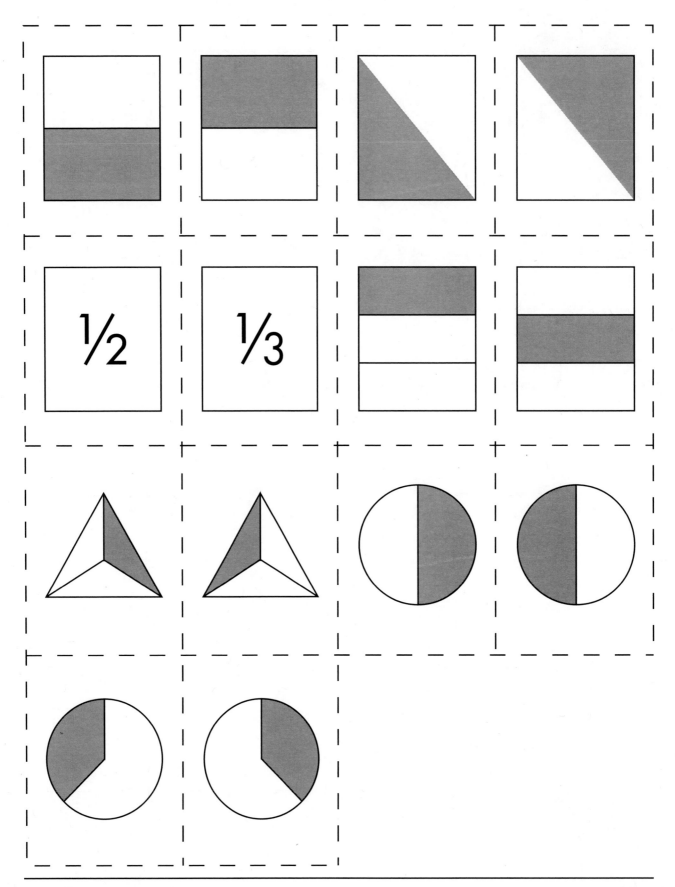

SHOW ME 1/2 GRIDS

SHOW ME 1/3 GRIDS

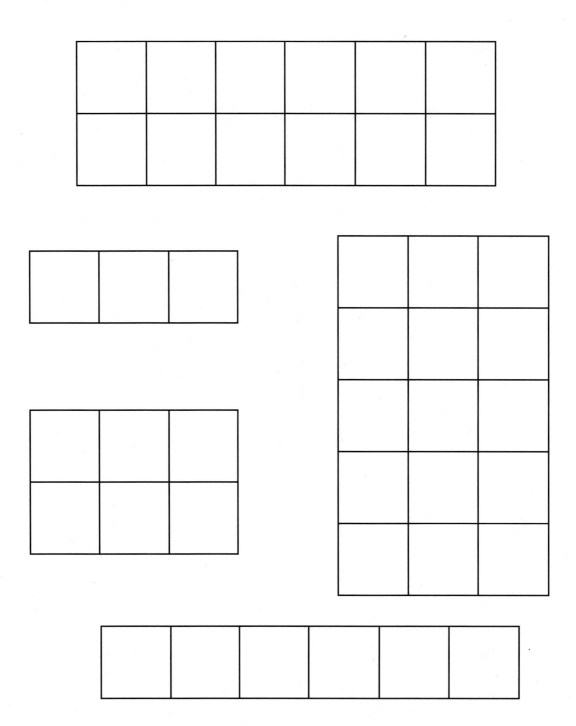

SYMMETRY GRIDS

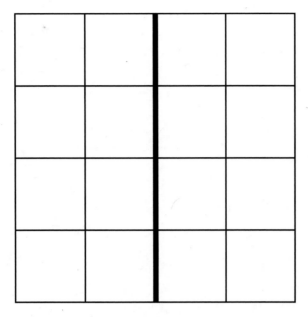

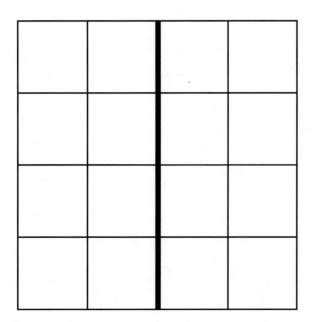

SHAPES

PAPER NUMBER LINES

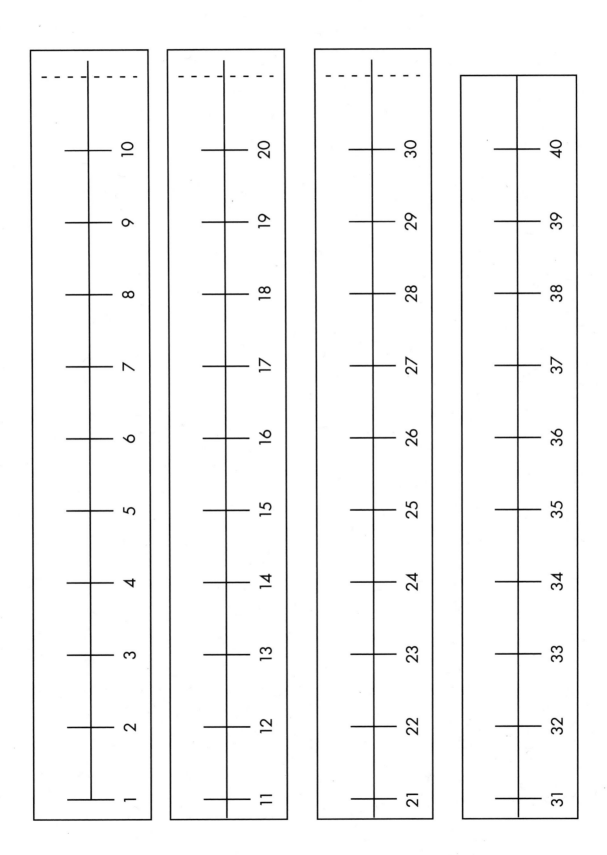

CLUES NUMBER LINES

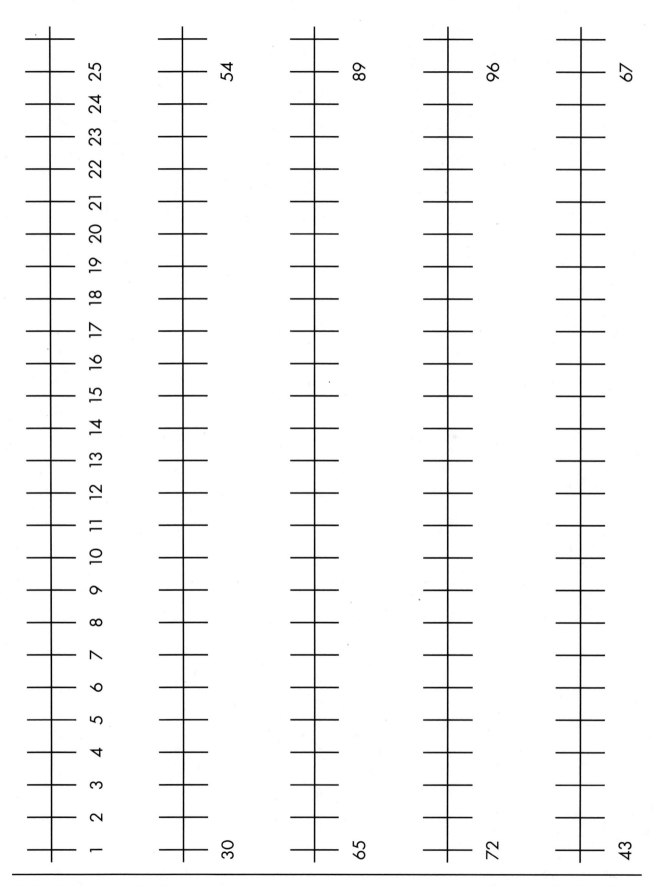